Unified Field Theory Solved:
Planetary Theory

By
Oliver Oyanadel

Copyright©2008 by Oliver Oyanadel.
First Published 107.5 millennia ago today
with Pre-Ice-Age Publications To:
Post-Ice-Age Publications,
whomever they may be at a Time.

All rights reserved. No parts of this book may be copied or utilized in anyway without explicit and lawful permission by our publishers nor without the full royalty payments rendered to the author and publishers.

One certainly cannot say there has never been an attempt to solve the Unified Field Theory. Most ultra-intelligent underachievers, maybe even, college stoner dropouts, become inspired by Einstein's Theory of Relativity just as I had, and when we learn that Einstein was never able to finish solving his Unified Field Theory before he died, most of us feel almost commissioned to solve it by sheer respect, let alone, gratitude, for having the unknown mysteries of the universe for us unlocked, or unshackled, as it were, an emancipation from the solitary confinement in which our comprehension had languished for eons, no longer all alone in the universe; Someone out there, was worthy of pondering the very essence of the universe–worthy of being granted the gift of finally spreading such intellectual wings of freedom.

As a rule, it is known no two independently different units should be added to each other when quantifying them together in mathematical physics, they usually should be multiplied, at least when being brought together as directly proportional to one another, and then of course divided when inversely proportional to one another. However, allow me to put aside all

Sinicism, and let's see if we can't enter this epiphany with childlike ambition for a whole new paradigm-shift in what we have considered to be so illusive for so long. And so not to bury the lead, this open mind will absolutely be necessary to unite Universal Gravitation with Electromagnetism, even though any child can see how they fit together just as the earth's continents once fit together as one land, Pangea. Especially when we see Newton's Law of Gravitation right next to Coulomb's Law. And since distance and velocity are units of length and time, they are exactly proportionally symmetrical all around the spectrum of all matter, be it subatomic or astronomic.

Newton's Law of Universal Gravitation:

$$F = \frac{Gm_1m_2}{r^2}$$

Where the attractive force between two bodies are directly proportional to the product of their masses and inversely proportional to the square of the distance between them. I do hate having to repeat the same old, "derivative dribble" as they say, but Newton really put it best, having discovered it to begin with.

And with,
Coulomb's Electrostatics:

$$F = \frac{C_1 q_1 q_2}{\kappa r^2}$$

The force depends on whether the medium between the two charges (denoted as q 1 and 2) are electrically conductive or insulating.

And for magnetic poles, each denoted with a, ρ,

$$F = \frac{C_2 \rho_1 \rho_2}{\mu r^2}$$

Coulomb's Electromagnetism depends on the

magnetic permeability of the field between them. Where the kappa symbol, K, is the dielectric constant and the, μ, is the magnetic permeability. So right there, we can see the difference between the medium of a conductive wire or cable, and the magnetic field of the radiation of some induction from the surrounding potential static charges.

When painting a still-life, an art student can begin by drawing only the negative spaces around all the objects instead of just focusing on the objects themselves. And this ends up bringing composition where there would otherwise only be a still, or cropped subject matter. And like substance and style must come together in equality. A magnetic field cannot exist without an electric charge. In fact, to say a magnetic pole is merely a fictional quantity just because one has no way of examining a single pole of positive or negative magnetism, would be as presumptuous as those whom assumed a single electrostatic charge of either positive or negative did not exist before we were able to literally clock one in a particle accelerator. And just as there can be no positive charge without a negative charge, there can be no magnetic fields without the matter that exists within the empty space in which the magnetic field resonates.

A charged Photon eventually decays into a proton, an

electron, a neutron (not exactly in that order), each decay into other particles expanding with the conservation of energy. And though these more stable particles won't begin to decay for up to 10^{31} years, there are other more cosmic rays that decay as fast as a 10^{-21} second or so. And considering just how fast they move through free space, to a particle's point of reference, it will have experienced 50 million years for every light year it has traveled before decaying to its final rest-mass.

Time slows down for whom approaches the velocity of light, which is approximately 186,000 miles per second. The earth is rotating at 1,000 miles per hour, and is revolving around the sun at 67,000 miles per hour. Multiply those together, and you have 67,000,000 miles per hour, which happens to be 186,000 miles per second, the velocity of light, (The velocity of light, which is denoted as c, is actually 67,200,000 miles per hour. The extra 200,000 miles per hour is explained in the mathematics of the following pages). Due to the big bang, we can include the velocity of the solar system itself, and this means we are actually moving at the speed of light squared (c^2). This is one of the reasons why our energy is at c squared, $E=mc^2$ (energy equals mass at the speed of light squared). The energy refers to our velocity on the planet earth. Most people instinctively want to disagree right here because it seems inconceivable to be experiencing the speed of light alone, much less the speed of light squared. But as Einstein said, time slows down as we approach the speed of light. So for each light-year the universe expands, we will have experienced 3 billion years. This is also one of the reasons we do not simply fly off the earth: because time has slowed down for us so much at the speed of light squared. The truth is we are traveling even faster. But we'll get back to that a little later.

Now, it is true nothing can ever really exceed the speed of light, but remember that all is relative. This is important. Because it all depends on which reference point you are comparing things to. The earth relative to the sun is moving at the speed of light. The sun relative to a fixed point in space, if there could be such a thing, is also just
moving at the speed of light. But since we are both traveling together and around each other, all at once, we are truly experiencing the speed of light squared (c^2 or c x c). And yet still, this makes the speed of light itself, from our perspective, actually traveling at the speed of light cubed, which is c to the third power (c^3). Now when you take all of that, and

consider the relative energy within all matter being at c as well (the speed of light as a constant of nature), since we are made up of molecules, which in turn have protons, neutrons and electrons within, and of course they are all traveling at the relative speed of light, and include all these energies together, we (the universe itself) are actually experiencing the speed of light to the sixth power! $(c + c^2 + c^3 = c^6)$. And this is the one and only reason we experience 3 billion years for every light year the universe expands. This means the neutronic mass value which created the big bang, was at the speed of light to the sixth power, squared. Which happens to be 2 to the 64th power, $BB=n2^{64}$ or $BB=n(c^6)^2$.

And we can prove all this by using the theory of relativity in its own structure with the Lorentz Transformation, because it works best in spherical structures, such as the decay of particles emitting different rainbow colors as they descend through the curvature of the atmosphere, much like a space capsule re-entering the atmosphere must glide it in at just the right angle or else it might burn up on re-entry.

Relativity uses the measurement of time to replace the measurement of length in common Euclidean math at the time, where any other measurement one uses to plug into the equation is replaced by the number, 1. Even though I did place a (1) in the equation as my measuring rod, it is still only going to represent time in this case, not only for congruency, but because, in part, it also proves the Unified Field Theory is now solved. And time being represented by the letter, t, will always be best understood simply as, 60 seconds.

$$t \bullet \dfrac{\dfrac{(c^6)^2}{c^2}}{\sqrt{1-\dfrac{(c^6)^2}{c^2}}} = c^2$$

60 seconds times the speed of light to the sixth power squared, divided by the speed of light squared, divided by the square root of the sum of 60 seconds minus the speed of light to the sixth power squared, divided by the speed of light squared, equals the speed of light squared. It all balances out. Furthermore, if we were to create a nuclear powered

spacecraft (or nuclear powered static-electricity anti-gravity spacecraft), where the whole outside metal frame would be an active static electric diode, the electric potential should be set to the speed of light to the 7th power squared (only as a numeric value in measurement). The parabolic shape of the emission of electrostatic radiation happens to be perfect in the Lorentz Transformation formula of the theory of relativity. Just like above, if you replace the formula with c^7 instead of c^6, it would equal c^3 instead of c^2. And as we have discovered before, the new relative speed of light is actually at c^3 **(c cubed or c to the 3rd power)**. And what this all means is that this spacecraft would be capable of traveling at the speed of light relative to us.

Now, once you travel at the speed of light, not only will time slow down for you, but your mass will also increase considerably. Our present velocity is what gives the earth its mass to begin with. So much so, that if the planet earth were actually an electron revolving around a nucleus, the velocity of our spacecraft, rocketing away at the speed of light relative to that electron (or faster than that electron), would bring it to the visible spectrum, making the planet electron a distant memory forever. How would one return? There would just be no way. As a matter of fact, the macroscopic is just as invisible to the microscopic as the microscopic is to the macroscopic. In fact, relativistically speaking, this is exactly what our planets and stars really are in the expanding universe from the big bang... That is to say, we actually live on subatomic particles; the same subatomic particles that erupted and decayed from the big bang at the speed of light to the sixth power squared.

The God I believe in, created the universe with a nuclear bomb with the neutronic mass value of 2^{64}. And He has that spacecraft, flying along side of us, invisible to us, but He of course, being the creator of all things, is capable of swooping in and out at His leisure, doing whatever He wills. And the rest, can all be discovered by the bible studies of the most wise, and the physics books already published today.

Time Dilation Of The Expanding Universe:

$\left.\begin{array}{l}\breve{r} = 1,000mph \\ \breve{R} = 67,000mph\end{array}\right\} = 67,200,000mph = c$

Solar System itself is in freefall at 200,000 mph due to the Big Bang and its constant velocity of c.

$\breve{r}$ = Rotation of the earth.
$\breve{R}$ = Revolution around the sun.
c = Velocity of light
t' = time dilation

BB = c,
$[\breve{r}$ and $\breve{R} = c$ and $BB = c] \rightarrow \acute{Y} = c^3$
$[\breve{r}$ and $\breve{R} = c] + [BB = c] + [E = mc^2] + [\acute{Y} = c^3] + c = c^6$
c^6 = the velocity of the expanding universe of matter.

$\sum_{0-1}^{r} Ly = 3$ billion yrs earth duration $\Big)$ if $c^2 = 50$ million years for each light year.

$$60 \cdot \frac{\frac{(c^6)^2}{c^2}}{\sqrt{60 - \frac{(c^6)^2}{c^2}}} = c^2$$

BB = Big Bang
Gamma = photon = γ
photon = particle of light
Gamma Prime = ÿ
ÿ = New Velocity of Light relative to $c^2 = c^3$
g = earth's gravity = $22mph/s^2 = 9.8m/s^2$
m/s = meters per second
mph/s = miles per hour per second
Ly = Light Year

The Universe Of Matter In Mathematical Form and The Creator's Map of the Universe:

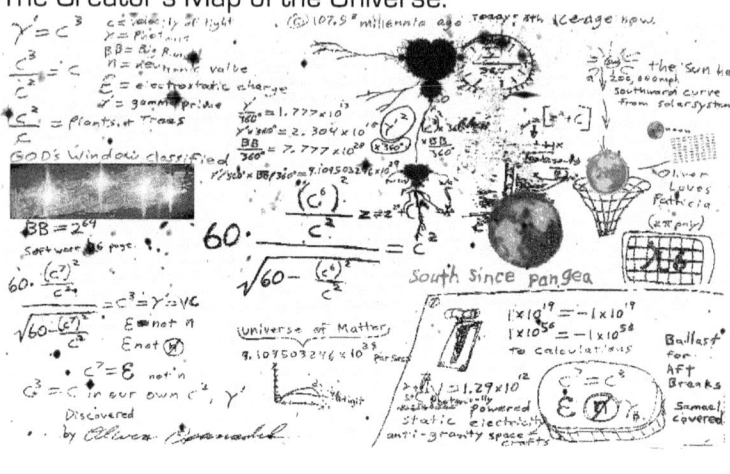

$$\gamma' = c^3 \quad c = \text{velocity of light}$$
$$n = \text{neutron}$$
$$\frac{c^3}{c^2} = c \quad \gamma = \text{photons}$$
$$\epsilon = \text{electrostatic charge}$$
$$\frac{c^2}{c} = \text{plants \& trees, the constant within all matter}$$
$$BB = \text{the Big Bang}$$

$$60 \cdot \dfrac{\dfrac{2^{64}}{3.5 \times 10^{10}}}{\sqrt{60 - \dfrac{2^{64}}{3.5 \times 10^{10}}}} = 3.5 \times 10^{10}$$

It has been written that the limits of the infinite have never been defined, but of course they have. They are defined as Three Hundred and Sixty Degrees of Uncertainty.

$$\frac{\gamma'}{360°} = 1.777 \times 10^{13}$$

γ' denotes the new speed of light relative to our own c^2. Where gamma, γ alone usually represents the photon, which is the quantum particle of light. And as you know, gamma radiation are high energy photons, known as gamma rays.

You may remember the actor who played the 70s version of David Banner in the Incredible Hulk had gotten cancer 30 years later, after shooting scene after scene of receiving gamma rays, consisting of green laser light beams scanning up and down his open eyes and straight into his retinas. This is why people should never play with laser lights. People need to stop shining ultra bright flashlights into people's faces for that matter.

$$\gamma' \times 360° = 2.304 \times 10^{18}$$

$$\frac{B \cdot B}{360°} = 7.777 \times 10^{28}$$

$$\left[\frac{\gamma'}{360°}\right] \times \left[\frac{B \cdot B}{360°}\right] = 9.109503246 \times 10^{29}$$

This seems to be the entire universe of matter in parsecs for the time being.

$$60 \cdot \frac{\frac{(c^6)^2}{c^2} z \leftrightharpoons z^2 + C}{\sqrt{60 - \frac{(c^6)^2}{c^2}}} = c^2 \textbf{ X } z \leftrightharpoons z^2 + C$$

Where C is a coordinate system, as apposed to the smaller c we are accustomed to representing the velocity of light and hopefully not to be confused too easily.

The z reiterated into $z^2 + C$ is obviously the M-set known as the Mandelbrot Set and employed here to represent the infinite objects in nature.

There are no straight lines in nature. In my analogy of the still-life artist, I compared the negative spaces of a composition to the magnetic field in open space, and relativity itself began by exceeding the limitations of Euclidean straight line Geometry by employing the axiom that there are no straight lines in nature. The

Fractal Geometry of the Mandelbrot Set is the perfect formula for all matter in nature, while the rest of the equation dealing with it's energy masses and units thereof, is how it all works in the field of open and outer space. This also achieves unification with all things.

$$60 \cdot \frac{\frac{(c^7)^2}{c^2}}{\sqrt{60 - \frac{(c^7)^2}{c^2}}} = c^3 = \gamma' = \text{velocity at c (relative to our own } c^2 \text{)}.$$

It is important to remember that this particular formula strictly refers to electrostatic charges and NOT any other type of subatomic matter, e.g. the result of such a mistake would be absolutely catastrophic, and should never be experimented on by anyone who is not a professional scientist. In other words, it only takes a small nuclear reactor to get your spacecraft up and flying. So these astronomical proportions resulting from these equations are in no way connected to it in any other way than the electric potential, and strictly as a numerical value on the electronic dial. Otherwise, some mad scientist would destroy the entire universe by creating another Big Bang right in the middle of our own home universe of matter. And I hope it goes without saying, that would be… Good Bye.

$$1x10^{19} = 1x10^{-19}$$
$$1x10^{56} = 1x10^{-56}$$

And this is how the astronomical units of Gravitational Mass relate to the particle units of Atomic Mass relative to electron volts, which reflects on how energy, being velocity, mass, direction, acceleration, time, and so forth, is in fact the stuff that *time itself* is made of.

The Frequency Of Matter:

$$6.4 \times 10^{15} = c^3 \rightarrow \acute{y}$$
$$4.1 \times 10^{31} = c^6$$
$$4.1 \times 10^{62} = (c^6)^2 \rightarrow 2^{64}$$
$$3.5 \times 10^{10} = c^2$$

$60 \times 4.1 = 246$
$246 / 3.96 = 62$
$3.96 \approx 4$
$4 \times g = 39.2$
$39.2^{24} \div c^2 \times 24 \times 365 = 2^{64}$ or $(c^6)^2$
$246^4 \div c^2 = -1 \times 10^{19} = mP$
$mP \div c^2 = -1 \times 10^{56} = Fe$

$10^{19} GeV = mP$
$10^{19} GeV \div c^2 = 2.857142857 \times 10^{28} GeV = 56 MeV$
$56 Mev = Fe$
$f = \frac{1}{t}$ The Frequency of all matter and reiterated infinitely within itself

Where mP or the Plank Mass is, at least in theory, the smallest possible particle, and Fe happens to be the Atomic mass of Iron.

If you divide the Plank mass by c^2 you end up with the atomic mass of iron (that is of course if you divide that answer by 1 million to make the transfer from giga-electron volts to mega-electron volts). Also, if the atomic mass of iron were positive instead of negative, it would equal the earth's gravitational mass, which is exactly the kind of symmetry that is synonymous with mathematical proof. Especially when seeking the unification of Electromagnetism with Universal Gravitation.

So Einstein's Unified Field Theory has been solved, and I call it, Planetary Theory.

Not String Theory...

Not Quantum Theory...

Planetary Theory
by
Oliver Oyanadel

So for a nuclear powered static electric anti-gravity spacecraft, the electric potential should be gauged at c^7, which, in this transformation equation would equal c^3, which would put the spacecraft's velocity at c in our own c^2.

If they haven't already, I'm asking for Merkaba Operation's Group to get working on it.

Operation: Almighty Please!

So now there's a new hope…

Unified Field Theory Solved.

www.ingramcontent.com/pod-product-compliance
Lightning Source LLC
Chambersburg PA
CBHW050308220526
45465CB00002B/872